NURSE V NURSE

My personal journey experiencing abuse in the nursing profession

Written by: A. Dhaie RN,BSN

Edited by W.H. Hochheiser

Acknowledgments

To my love, words cannot express how deeply I am grateful for all of those years you stood by me. Thank you from everything that I have. To MSM you are the other half of my heart. Thank you to S and k. You pulled me out of my darkest hours. I cannot say thank you enough. To J and L, without you, this would still be on the shelf. Thank you to Pixie who first asked that question that made me take pen to paper and to E. Corda RN, BSN. I was honored and blessed to have worked with you for so many years. You are the nurse that I always wanted to be. You are fearless, personally accountable, and your nursing skills have been and always will be unparalleled.

Forward

Many times I sat down to tell this story and stopped. I thought maybe what I went through wasn't really bad. I thought maybe if I had been stronger, better, or smarter, it would have been different. I thought maybe in some way, it had been my fault. When I reached out to nurses who are still working where I had been, I realized what I went through wasn't just about me. If it had been, those nurses wouldn't still be going home crying, scared, and insecure and questioning their abilities. Nurses with ten, fifteen, even twenty years more experience than I have are leaving work every day feeling that they have failed. I finally realized I am writing this for them, as much as for me. I want to be clear. I am not saying that I haven't made my fair share of mistakes. I am not perfect. I am not saying I have always had the best response or was the best I could be. I am not saying I was always above reproach or correction. What I am saying is this: what I went through left me broken, questioning, and trying to put the pieces of my life and myself back together again.

And I am not alone.

How a unit works

To understand what happened, you need to have an understanding of how a medical/surgical unit in a hospital typically works. There may be some things here that are specific to where I worked but, in general, this is how things go. When a patient needs to come to any floor, either from the emergency room, or a transfer from another floor, the charge nurse will get a call from either the "bed board" or the house supervisor. The people on the bed board are responsible for knowing where every person is in the hospital building. Whether a patient is waiting in the emergency room, or in the pre-surgical area before a procedure, the bed board knows. They track every patient every minute of their stay. From intake to discharge the bed board is responsible for tracking the flow of patients through the hospital. But it is important to note here that the members of the bed board are not medical people; they are data people. They get the information and they pass it on.

The house supervisor is the person responsible for the flow of patients through the hospital. At night and on the weekends when upper management are not physically present in the building, the house supervisor is considered the boss. This makes the chain of command drastically different from day shift to night shift. On the day shift the chain of command goes from change nurse to unit manager to house supervisor to director to administration, and up from there (depending on the hospital's structure). On nights and weekends the chain of command is from the charge nurse to house supervisor to administrator on call.

Acute problems on weekdays are normally solved by the unit manager with the house supervisor called in if the unit manager isn't available. On nights and weekends problems go right from the change nurse to the house supervisor. On a medical/surgical unit, this means that house supervisor has authority over the charge nurses and floor nurses. In general, house supervisors are nurses. Typically, they have extensive nursing experience, but not always. They are higher in the hierarchy than the bed board. The bed board can *ask* a floor to take a patient while the house supervisor can *force* a floor to take a patient. House supervisors are held more personally responsible for the flow of the hospital. For example, house supervisors are responsible for the time it takes for a patient to go from the emergency room to the floor. If the floors are unable to take a patient due to staffing or bed availability (that is, every bed has a patient in it), the house supervisors must answer for that.

Let's say that there is a patient on the medical floor that needs to go to the ICU and there is only one bed left in the ICU. The house supervisor is responsible for negotiating what happens next. Does she try to keep the bed open in case they need it, or does she move my patient up and risk closing the ICU? These are the types of decisions that the house supervisor is responsible for making.

The charge nurses, on the other hand, are in charge of their floors. I worked as a charge nurse on a fifty-bed medical/surgical unit that also had oncology patients. One of my biggest responsibilities was placing patients who needed a bed on our floor from the emergency room, or from another floor. When a patient needed a bed and the bed board or the house supervisor believed our floor was the best fit, I was contacted and given information about the patient.

Usually this included the patients name, age, gender, diagnosis, special needs (like one to one care, needed to be placed close to the nurse's station, or needed a private room), and name of the accepting physician. If it sounded like they qualified for our unit, then my next step was to determine which nurse was in a position to take a new patient safely, based on the needs of their current patient load.

Then I checked to make sure the room and bed were clean and ready to receive a patient. Next I would talk to the tech and let them know they had a patient coming. Techs, or nursing aids, are the people on medical floors who take vital signs, do baths, and change linens. They have less education than nurses but make no mistake, a skilled tech can make the difference between a live patient and a dead one. After I made sure that the nurse, tech, and bed were ready to receive a patient, I called back the bed board or the house supervisor and gave them the bed number. Then the receiving nurse or the charge nurse (in this case me) would "take report" (that is receive the relevant information) about the patient from the handoff nurse.

Finally, the patient was brought to the floor by transport, a tech, or a nurse. When this whole process went smoothly, it took less than twenty minutes. There is tremendous pressure from upper management to get patients (who have already been seen by a doctor) from the emergency room to a floor within 30 minutes. This is considered "best practice" in medicine and is quickly becoming a goal for every hospital in the nation. Everyone is under pressure to get everything done within that time frame for every patient. However, things rarely go that smoothly. To hit that thirty minute target, we needed a nurse and a tech who were able to take the patient, the room and the bed had to be clean, and the patient had to be appropriate for our floor. Without this perfect storm, it would take much, much longer.

If the bed was dirty and housekeeping was free and able to get to it right away, it added about fifteen to twenty minutes to the process. If the bed was dirty and housekeeping was busy or shorthanded, it would take another thirty to forty minutes to even get to the bed, and then another fifteen to twenty minutes to get everything clean. If the previous patient in that room was in isolation, it would need a special kind of cleaning and could take much longer. This was NOT because housekeepers weren't doing their jobs but because of the sheer amount of work involved with getting a room sanitary after it had housed a patient who required isolation due to a contagion.

If the nurse was able to safely receive a new patient, he or she could take them right away. More often, nurses needed another fifteen to thirty minutes to be in a place to SAFELY accept a patient into their care. "Safely" accepting a patient means that the nurse had the time to go see that patient when they came to the floor, and assess them properly, without having the needs of their other patients interrupted, and without making the new patient wait for hours for initial assessment. Ideally, a receiving nurse would assess a new patient as soon as they hit the floor *for that patient's safety.* A nurses list of emergent duties (things that need to be done right NOW) could be so long that receiving a new patient at any given time could have been effectively unsafe for both the patient and the nurse.

There were also questions about whether the patient was truly a medical patient or met criteria to be on another floor. Every floor in the hospital had a different set of requirements or criteria that a patient had to meet to be placed there. Sometimes it was obvious; surgical patients go to the surgical floor, oncology patients go to the oncology floor. But, other times, it has to do with more specific criteria. Those multitude of criteria are what separate a medical patient from a step down patient from an ICU patient.

The general rule was that a medical patient could only be on a medical floor if they didn't require an intervention more frequently than every two hours. For example, if a diabetic patient required their blood sugar to be checked every two hours, then they weren't appropriate for the medical floor. The time required to complete those interventions qualified that patient for a higher level of care. A step down nurse with a nurse to patient ratio of five patients would be able to do something for each of their patients every two hours, but a medical nurse with a load of seven, eight, or nine patients would not be able to cover those needs.

As a general rule, each medical nurse in the hospital had the highest patient load of any specialty. This is because for a patient to qualify to be on a medical floor they had to be considered stable. It goes like this: an intensive care nurse would have only one to two patients because of the intensity of patient's needs, in other words how often the patient would need something done and how complex that particular intervention was. One intensive care patient required more time than one step down patient. Step down nurses would have five patients apiece because their patients need attention more often and with more complexity that that of a medical patient. Since medical patents are considered stable and typically require less personal, time consuming, and complex interventions, the medical nurses get a higher nurse to patient work load.

Patient load is referred to as our "grid". At our hospital, our grid for the medical floor was based on seven patients per nurse at night and six patients per nurse during that day. The "grid' is the financial back bone of the hospital. The highest expense a hospital pays out is for nursing. The two floors that make the most amount of money are the surgical floor and the medical floor. Let's say the grid is seven patients to one nurse on the medical floor. This means, at that ratio, the hospital makes a profit. When the medical nurse has six patients per nurse, the hospital loses money. And as you can imagine, when the medical nurse has eight patients per nurse, the hospital makes more than the budgeted profit. This is extremely simplified of course, but you see where this is headed.

As long as I have been doing this, no one has ever answered my question as to who actually makes the decision about the grid, patient load, and profit. They told me that it was not the unit manager or even some of the upper management. But the decision to "break" the grid and give the nurses more patients than the grid dictates tends to be made by the house supervisor, unit manager, and chief nursing officer. At least this is what I had experienced. On a medical floor, because the nurse/patient loads were so high, nurses only had so much time to care for each patient. If a nurse received a patient that required intervention (direct involvement) more often than every four hours (such as lab work, mediation or bedside care) then something else would simply NOT get done. This created an unsafe situation for both the nurse and the patient, i.e. medications were missed, labs weren't done or weren't reviewed in a timely manner, procedures were missed, and transfusions remained undone. There was also a grid for techs, but that was consistently ignored if the nurses were able to receive patients.

If a patient needed to be sent off the unit, for example, for complications like respiratory failure which requires a ventilator, cardiac failure that required certain types of medications, having seizures, uncontrolled blood pressure, or, some other process that was necessary to receive an intervention more than every four hours, I would start the process of moving them to another floor by calling the bed board and requesting a bed. Then the current nurse would call report to the receiving nurse. (It was also not uncommon for me, as the charge nurse, to get into the chart and call that report to the receiving nurse.)

On a medical unit, the patient was always escorted off the floor by a nurse. If everything is going smoothly, the patient could be transferred off the floor as quickly as fifteen to thirty minutes. If there were problems (such as inadequate staff on the receiving unit, no clean beds, or the charge nurse was tied up in an emergency situation and couldn't assign a bed), it could take four or more hours to transfer a patient. If the patient requiring transfer was unstable, the patient's nurse would IDEALLY not take any new admissions (and should have a lot of help with her other patients) so that they could spend critical time with the unstable patient prior to transfer. Sometimes this situation required calling in the "react team". The react team was a group of nurses who had ICU (intensive care unit) experience. They helped the floors assess patients, communicated with doctors, and assisted in complex situations. Sometimes a react nurse had to stay with the patient until they were transferred off the floor. However, when this happened, it meant ALL other react duties were also put on hold until that patient had transferred.

What I really want you to understand is this: it didn't (and still doesn't) take much for three of four people to become completely engaged in one patient's care. When this happened, the other patients assigned to the primary nurse were not seen in a timely manner. Call lights couldn't be answered by a tech already involved in a complex patient care. The charge nurse couldn't take information on new patients and place them on the unit, and (if the react team was involved) other critical patients weren't seen. If the transfer was complicated by a lack of clean beds, low staffing, or the receiving floor's charge nurse was in the middle of caring for her own critical patients, that transfer could take hours. However long it took to sort out all of those factors, the three or four people who were responsible for the one critical patient were tied up for the entire amount of time.

If the house supervisor was still available, then they could try to speed up a transfer, but that wasn't always the case. This is why it is so important that every patient got placed on the right floor from the onset. The amount of work that it took to care for and transfer a critical patient off of a medical floor was onerous. Moving a patient off a medical floor and onto a unit could take hours and involve many people from multiple departments. In the best case scenario, which means that the people involved are organized, have excellent time management skills, and that nothing unexpected comes up in the process, it can happen over fifteen to twenty minutes. A less optimal, but more likely, scenario can easily take one to five hours. Remember that this is a transfer from floor to floor, which is different than a transfer from the emergency room to the floor, where thirty minutes is the ideal.

Having a base understanding of the kind of time and multi-personnel effort that it takes to properly place a patient and then move him or her if necessary, will help you understand why the nurses are under the kind of pressure that they are. Every patient in the emergency room is put on a timer, the times starts as soon as they come into the emergency room until they see the doctor. It continues from the time that they see a doctor and ends as soon as they come to the floor. Every patient that needs to be transferred to a higher level of care is on a kind of clock but not as literal as the one in the emergency room. If the patient attempts to sue the hospital for malpractice or injury, it can be argued that the time that it took to transfer them to another floor is unreasonable, thereby leading to further injury or death. So, every patient in the hospital is on the clock so to speak, and for the medical staff, it is the race against that clock that runs the hospital. In addition to all of the external issues that impact the transfers and transfer times of patients, other factors that must be taken into consideration are the personalities and behaviors of both the patients and the staff involved.

Types of Behavior

I have been in nursing for 25 years, ten of those in a hospital. Over the last few years the concepts of "lateral violence"," that is to say nurse-to-nurse violence, and bullying have become hot topics in the nursing field. However, just because we have begun to talk about it and have some names for what is happening, it doesn't mean we have the ability to mitigate these behaviors or to actually deal with them in real time. Trainings that cover these issues tend to be more focused on "don't do it" rather than "here is how you address it in the moment, when it is happening." Additionally, doctors who are abusive to staff but bring in significant money, or who work the unpopular shifts that no other doctor wants to work, will remain employed no matter what the complaints are, even if those complaints include physical violence or inappropriate interactions.

The same is true for nurses. I have seen nurses keep their jobs after repeated complaints about openly masking racist, classist, homophobic, or religiously biased statements to patients and families. I have seen nurses keep their jobs while allowing their biases and bigotry to influence their treatment decisions, quality of care they provide, and which staff they support. Such complaints are generally ignored regardless of whether they come from staff, patients, or family members. One unfortunate reality is that people are sometimes bigots and that does not separate itself when that person is a nurse. The other unfortunate reality is that negative behaviors of nurses and techs are routinely excused. Legitimate concerns about quality of care combined with a nurse's disparaging comments are regularly brushed off with excuses from management like "they were only kidding" or "they have always been that way" or "no one else seems to have problem with it," and any number of other rationalizations which allow the behaviors to continue.

Bullying and abusive behavior come from any source. In nursing it isn't uncommon for nurses to experience it from EVERY angle throughout the course of a single shift. To give you an idea of what I am talking about, here is a list of common behaviors nurses encounter every day. I organized these behaviors according to those who exhibit them. These are behaviors that I have experienced personally and I have seen happen to others consistently over the years. Even when reported, these behaviors are allowed to continue with little or no intervention from management.

Doctors

1) Not answering calls or pages. *I have been in emergencies where the doctor won't answer his or her pager, phone, or overhead page, and can't be found on the premises. The react crew had to transfer the patient and assume care for hours until the doctor was located.*

2) Talking too fast to be understood, then belittling the nurse for not understanding.

3) Throwing charts or other objects, sometime at nurses. *Surgeons are notorious for being physical with their staff.*

4) Ignoring requests (i.e. clarification on written orders, ect.) and then blaming the nurse. *I have heard a lot of doctors justify this by saying "You should have caught it the first time".*

5) Speaking loudly and badly about a nurse in a public area where they know they will be overheard.

6) Refusing to answer care related questions, or pretending not to hear them, even when the nurse is right next to them. *I have been right next to doctors, asked for their attention and have had them ignore me outright. I was left standing, waiting for their attention while they texted, or played games on their phones.*

7) Saying "do whatever you want" and then hanging up on the nurse. *I had a patient who was actively seizing, most likely due to coming off of alcohol. The secretary paged the doctor and after multiple pages, the doctor finally answered. I was able to tell her the patient's name and date of birth when she interrupted and said, "do whatever you want" and hung up.*

House supervisor/Upper management

1) Ignoring the grid

2) Ignoring the qualifying reasons why patients need to be on a medical floor.

3) Yelling at and insulting nurses in public areas, in front of other patients and other staff. *When I was first at the hospital it was never done. When the new chief of nursing officer took over she would personally come down to the floor during the day and yell at the staff. This was then considered acceptable behavior and was quickly picked up by those in her ranks.*

4) Falsely accusing or humiliating the charge nurse or the floor nurse, sometimes privately, but also sometimes in front of his or her staff and or patients.

5) Creating no-win confrontations in front of others, on the floor or in meetings, and refusing to listen to or consider explanations/circumstances. *I have seen this played out when upper management pretends to give the staff a choice with no intention of accepting the staff's input. It is an effective way of disempowering people and doing it quickly. One instance that sticks out in my mind was a charge nurse meeting where the chief nursing officer asked the charge nurses what we thought about changing the shift start and end time. The group was adamantly against the change. The chief nursing officer said "try it for a month" and "we will talk about it at the next charge nurse meeting." A month later the group met again and was still adamantly against the change with good reason. The change was found to be, across the board, disruptive to the staff overall. But the chief nursing officer implemented it anyway. We never really had a choice.*

6) Accusing nurses of being incompetent, insubordinate, or otherwise "difficult to work with" when they question taking a patient that is not qualified for their floor due to higher level of care needed. *I saw this time and again. The doctor placed the order for a higher level of care and the house supervisor would berate the charge nurse or the staff nurse because of the patient needing a higher level of care.*

7) Personally accusing the nurses for low staffing levels or having no beds available.

8) Lying outright to upper management regarding the floor's status, the staffing, the care of patients, or the nurse's response. *The first time I ran into this I was shocked. A house supervisor was pushing for beds and telling me that the emergency room was overflowing and they had to move those patients out quickly. We had been pushing all night, taking patient after patient. I went to the emergency room and I saw only one patient. The waiting room was completely empty and the emergency room staff was being sent home for low census.*

9) Insisting that unsafe patients be placed in the unit, overriding the doctors' orders and the charge nurse in order to do so. *If the house supervisor or the chief nursing officer wants a patient out of the emergency room, they dump them on the medical floor, even if they should be placed in the ICU. Hospital emergency room wait times are tracked and reported nationally but wait times are not tracked for the floor-to-floor transfers. What this means is floor transfers do not affect the national quality measures. It quickly becomes a matter of*

"get the patient out of the emergency room any way possible, then deal with it later."

10)   Placing patients on the floor with full knowledge that there are not enough nurses on the next shift to safely cover their care. *This happened routinely. The next shift was considered "later" even if "later" was in ten minutes. I was made to take patients knowing that the day shift would have nine patients apiece instead of the grid of six patients to one nurse.*

11)   Asking for input and opinions with no intention of utilizing that information, except to label those who gave their opinion as troublemakers who don't support the organization.

Nurses

1) Refusing to help, often claiming to be too busy when they are obviously not engaged. *I found this to be very common. Those nurses who are too busy to help others always seem to need and expect help for themselves.*

2) Refusing to show up for a code. *Yes, I just said that . I have been in situations where nurses refuse to respond to a situation where the patient was in cardiac arrest or respiratory distress.*

3) Making up gossip or starting rumors that are professionally damaging.

4) Not turning in paperwork or not completing assignments for hours and having to be chased down for it, thus delaying transfers or other elements of patient care. *I worked with a nurse who had been at the hospital for about fifteen years before I got there. She took twelve hours to do an admission and every time it was turned in incomplete. Hospital policy was that admissions need to be completed within two hours. I rarely held nurses to that, because I didn't have to. If the night was busy and the nurses turned them in within three or four hours, that was fine. For other nurses this was the exception and not the rule. For this person it was twelve hours, and incomplete, every time. I talked to her, then reported her over and over. She would get better for one day and then the next day be right back at taking twelve hours. It was exhausting chasing her down every day. So I had to measure out where I wanted to put my energy, time and effort. Chasing her down, or assessing patients in need?*

5) Publicly giving gifts to all but a select few coworkers. *I have seen what this does to the left out people. It creates self-doubt, anger, and sometimes tears.*

6) Not discussing issues with a coworker but reporting everything about that person (no matter how insignificant) to upper management. *This*

narrows the focus on someone, making all their mistakes seem bigger than they really are. I have seen this over and over. It does two things: one, takes the spotlight off of what the reporting nurse is or isn't doing and, two, it focuses attention on someone else.

7) Intentionally making the giving and receiving of report an arduous and confrontational experience.

8) Nitpicking the work of others while often doing the exact same things themselves.

9) Saying things to get you to agree with them then going around and saying you said it in the first place. *For example:*

Nurse A: Sally is really slow.
Nurse B: Yes, she is.
Nurse A to Nurse C: OMG, did you hear that nurse B thinks Sally is really slow?
It took a while for me to pick up on this, but I have seen and heard it in action. It is very effective team splitting and does wonders for creating animosity.

10) Instigating discord by initiating conversations about situations that are beyond staff control (like grids or raises) to get and keep people agitated and upset. *I have watched this in action. One nurse was sitting with two others. She wanted to count down how many nurses have quit in the last six months and why. The other two nurses didn't*

want to participate, but she insisted. Before they knew it, the negativity in that corner of the room was palpable. The other two nurses' moods turned sour, and the nurse that initiated the conversation sat back and enjoyed the complaining.

11) Talking badly about or blaming issues on other nurses or techs in front of patients, or right outside of patient rooms, loud enough to be overheard.

12) Refusing to do a task or saying that they don't know how to do a task, so they have to be told how to do it every time it is needed. *When I have experienced this, it is with the same people over and over. I would go through a procedure with a nurse one day, or sometimes several times in one day only to have him or her come right back to me the next day stating that he or she did not know how to do it, as well as asked to be shown again.*

Techs

1)   Refusing to answer their phones or pagers.

2)   Refusing to answer their call lights. *This is truly an epidemic, partly because they are horrifically understaffed and partly because the calls from the patients can come nonstop for twelve hours.*

Remember that techs also have a grid that guides their tech to patient ratio, and more often than not that grid is disregarded.

3) When asked to do something, answering with "I'm busy" then hanging up or walking away. *When this would happen I was usually asked by the nurse to talk to the tech or to ask the tech to do the task. It would get so bad that the nurse would rather double their own workload then ask the tech to do something. The nurse didn't have the time or energy that it took to fight the tech to get the simple things done, so he or she got behind on their own work instead.*

4) Not answering a request, and then snapping at the nurse with words like "I heard you." When the nurse asks for a response.

5) Hiding

6) Not responding to a code or an emergency situation. *I have seen techs not respond to a dying patient just because they didn't like the nurse that was taking care of that patient.*

7) Responding to directions by saying that they don't know how to do a task that is a routine part of their job.

8) Arguing or excessively questioning instructions so every request becomes a "will they or won't they" scenario.

9) Not passing along relevant information to the nurse. *This can happen for many reasons from not liking the nurse, being too intimidated by a nurse to communicate with him or her, or simply not choosing to. Information could include extremely high or low blood pressures, blood sugars, or significant changes in level of consciousness of a patient. Occasionally there is a tech that says that they "didn't know" that they were supposed to pass on information. This always dumbfounded me because this is common routine procedure. What I found later was that techs that used the excuse "I didn't know" typically were disengaged from their jobs or had experiences with that nurse that were unpleasant and hostile. Their "I didn't know" was their way of avoiding uncomfortable conversations.*

10) Leaving the floor in groups so that they are not enough techs, or sometimes no techs, on the floor.

Secretaries

1) Not passing along relevant information to the medical staff. This can happen for the same reasons listed above for techs.

2) Not following up on doctors' orders, or writing off orders as complete without actually doing them.

3) Throwing charts. (Yes, physically throwing things.) *I remember when this happened to me. The day secretary came in as I was in a code, and*

a new admit had just arrived. I had not had the chance to call the bed board to let them know that the new patient had gotten on the floor. When I sat down after the code to call the bed board, the phone rang, it was bed board wanting to know if that patient had arrived. The day secretary answered it, attempted to answer questions and got so angry that she threw a chart. Managements response to her behavior was "that's just who she is."

4) Not answering the phone or the call lights.

5) Not getting paperwork to medical staff.

6) Not following through on requests for outgoing calls, paperwork, or other needs.

7) Ignoring requests or questions. Or ignoring requests from specific people but doing it the right way for others.

Patients

1) Belittling the nurse for things that the nurse cannot control. *Usually in this situation, when it became abusive, I, the charge nurse, would go in. The verbal and emotional abuse heaped on the nursing staff by*

patients and patient families is unreal, and a part of daily nursing life. Because nurses are held accountable for the quality for the "patient experience" we are discouraged from stopping abuse when it happened. We could be written up or otherwise reprimanded if we spoke up. Nurses were often counseled that we were responsible for a patient's behaviors.

2) Calling the staff derogatory names.

3) Yelling at and cursing at staff.

4) Physically assaulting staff.

5) Threatening staff. "I will get you fired/have your job." *This is a common everyday threat. I heard it every two out of the three nights that I worked, and when I didn't hear it my staff did.*

6) Saying things like "I pay your salary," while insisting that they receive something that they are not medically cleared for. Or demanding to be moved up in the line for tests or procedures.

7) Taking pictures or video without the staff's consent.

8) Refusing to consent to care and then blaming the staff for the outcome.

9) Insulting staff because of their age, appearance, race, gender, perceived intellect, or accent. *I would often swap out that staff*

member and then make a note for the next shift to not give that patient a nurse who is (fill in the blank here). Although many nurses are prepared for this, it was often the inappropriate sexual commentary or advances from patients that got to the nurses the most. For a young or new nurse this can be very complex as they may not know how they're expected to react in the face of a verbal assault or an inappropriate advance.

10) Accusing the nurse of purposefully withholding medications or care regardless of doctor orders or emergency situations. *This is very common with pain and anxiety medications. Even if the doctor has specifically written instructions for NO more pain or anxiety medications or their patient's nurse is an emergency situation and could not give the mediation right away.*

I considered not including the section on patients because I have had a number of incredible, kind, thoughtful patients throughout my years as a nurse. I know some patients had legitimate issues. However, driven by patient satisfaction surveys and subscribing to the idea that the patient is always right, nurses are required to put up with abuse (up to and including physical violence) and are often blamed for that abuse. Bullying and bad behavior from patients and families is a part of the DAILY nursing experience, and it is time we admit it.

My Story

(Please note that the names have been changed.)

The first five years I worked at the hospital were "normal". I worked nights. I was thorough and organized when it came to beds and nurses. I stayed in touch with the floor nurses throughout their shift and made it a point to know what was going on with each person. I did bed rounds early in my shift. This included physically looking at every open bed and taking note on its status. (Private room, male or female, was it clean or dirty?). Doing so allowed me to have the entire list of available beds at my fingertips and that, coupled with the knowledge of where the nurses were in their patient loads, meant that I could give out beds almost immediately, sometimes at the moment they were requested.

As a charge nurse, I did not have a patient load (which is very unusual in the hospital environment) but I had more than enough work to make up for it. When I first came to the hospital, the night nurses took eight patients apiece, with the day shift taking seven patients per nurse. Over the next year this was changed to seven patients per nurse at night and six patients per nurse during the day. I got along well with the house supervisors all except one whom I will call May. She was shorter on patience than the others, but I figured that she was just a little high strung and I would need to be on my toes when she worked. There were times when I would get a request for a bed from bed board and then call their emergency department and ask for more information, trying to assess if the patient qualified for our floor. (Remember bed board were not medical people). May would then call me and shortly ask, "Why are you refusing to take this patient?" I would calmly try to explain that I wasn't refusing, but was trying to get more information for proper placement. The same thing would happen as I was working with the nurses to make sure the patient could come to the floor safely, trying to get beds clean, in the middle of a procedure, or other emergency situation.

None of these explanations or situations seemed to matter to her. To May, it was always that I was refusing to take a patient. Since these incidents were minimal, we didn't work together on a regular basis, and I received a lot of positive feedback from the other house supervisors, I didn't really put much stock in her behavior. I just knew when she was working, it was more than likely going to be an unpleasant night.

From time to time May would cross a boundary. One night when I received the information on a patient from bed board, I recognized the patients name. She had been on our floor many times before, and she was always in need of a sitter when she was hospitalized. For safety reasons, she required a tech that would sit at her bedside and attend to only her. I called the emergency department to get information that bed board had not been given, a common practice for me. As I was checking with the doctor, May called me and accused me of refusing to take the patient. She demanded a bed number from me, ignoring my explanation of what I was doing and why I was doing it. She was so hostile and accusatory that with that incident I finally started reporting her. After that, things got more out of hand.

On more than one occasion, when I was with a doctor performing a bedside procedure, May would call demanding that I leave what I was doing and come to the phone. When I could not, she escalated, becoming more hostile. May would scream at the secretary to come get me. I would be at the bedside assisting a doctor with a procedure on a patient and the shaken secretary would be in the hallway nervously telling me that I was ordered to stop what I was doing and come to the phone. Afterwards I would try to explain myself, but May would talk over me and cut me off, accusing me of being uncooperative and then either hanging up on me or continuing to yell at the secretary.

Sometime I would report her to my supervisor (which was my unit manager, whom I will call Sue), and sometimes I wouldn't. I reported her in the past, yet there had been no change in her behavior and no response to me regarding my concerns. I would base my decision to report May (or not) on the intensity of the incident, where I was in my work week (if it was the last day of my week, I would tend to let it go) and, quite frankly, how emotionally and physically tired I was after my shift. If I was utterly exhausted and falling asleep on the toilet, I simply did not have the energy to write a long email, or even a short one. I had learned that if I called Sue, she would ask me to write her an email. All of these factors played a part in whether I reported her behavior or just walked away and tried to enjoy my days off.

To be completely honest, I downplayed her actions in my head. I justified what she was doing by telling myself "she has a tough job" or "she was having a bad night" or "it wasn't that bad, I am just over thinking this." I excused her behavior and worked to convince myself that I was overreacting. I think I really believed that if she knew how she was coming off and the effect it had on others, she would behave differently. I know now that I was wrong.

At the time, I felt like I was banking my time, creating a sort of mental debt almost, where she would surely see that the nurses and I worked really hard and feel that she owed us better behavior. I thought if I let the other stuff go, when I really needed her to, she would *have to* listen. I felt as if somehow, I was earning my stripes with her. I figured she would eventually see I am a good worker and (while maybe not backing off on the nightly stuff) on the nights the nurses were really in a bad situation, she would give us a little room to breathe. Maybe then, I thought, she would not make us take a patient right away or would otherwise be more helpful or considerate. If I could take her abuse, I told myself, the nurses I worked with wouldn't have to. If I could just "prove' myself to her then, well, I don't know what. Maybe it would be better. Like any victim of abuse, it made sense in my head at the time.

Over time May became more unreasonable and hostile. She would lie about what I said to and about other charge nurses. I would catch her in these lies by communicating with the other nurses myself. For example, one night I needed to transfer a patient to a different unit. It wasn't acute (time sensitive) so I spoke to the charge nurse on the receiving unit, and we came up with plan based on her staffing. However, May wanted the patient transferred right away (she had not seen the patient, did not assess the patient, and had not read the chart) and told the other charge I said transfer needed to happen now, and that the other charge nurse was refusing to take the patient. I never said anything like that. What could have been a really big, accusatory, and hostile mess was diffused because the other charge and I spoke directly to each other and worked it out. Fortunately, the other charge and I had worked together for a long time and had a good deal of respect for each other. Had that not been the case, this easily could have created some difficult emotions that would have affected relationships between the floors and could have impaired patient care going forward.

Another continuous issue between May and me, was observing staffing requirements. When May was working it didn't matter if I had seven patients and every other nurse on the floor also had seven patients. May would insist on nurses being pushed to eight patients. Even if the next shift staffing was low, May would still fill our floor. This would leave the day nurses with seven, eight, and sometimes nine patients apiece, including a full count of patients for the charge nurse. May would say it didn't matter what tomorrow looked like; we had to place the patients today, even if tomorrow was literally minutes away.

House supervisors all have different reasoning for the decisions that they make. I believe that for May, her decisions were to make her look good. If the emergency room was emptied early and she sent staff home saving the hospital money on productivity and increasing profits, of course she was going to look good. So deciding to overload the staff made productivity look good, or rather made her look good.

Making nurses take a higher patient load than the grid dictates has a direct and immediate effect on the staff: increased burnout, turnover, negativity, decreased morale, and decreased effectiveness in nurses. When staffing was consistently bad, I noticed over time that the same conversations between staff nurses and techs was happening again and again. It involved how they as medical professionals felt ineffective, like they were not able to do their jobs to the best of their ability; how the shift was about surviving and not really serving their patients. In essence it eroded their nursing pride and self-image in their jobs and identities. For many nurses, being a nurse isn't about having a job as a nurse. It is a matter of pride that they are "A Nurse" and that they can be the best nurse that their patients deserve. Not having that time to watch patients' needs eroded that at its core.

If May thought I wasn't moving fast enough, she would threaten to call the administrator on call to complain about me. She would stand right in the nurse's station and threaten to do it in front of the entire staff. This is a huge deal: the administrator on call was usually the chief nursing officer or the chief executive officer. Being on their radar meant you were in serious trouble. It didn't matter what time it was, or if I was running into significant staffing or patient problems; she would threaten me and then would make the call if she didn't get her way. What this meant was that instead of me being able to get my work done (my actual work: assessing patients, assisting nurses, etc.) I would be stuck in the nurse's station having to explain myself, my nurses, my staffing, and or my situation to May and the administrator on call for thirty to forty minutes. After that I had to catch up on the work that wasn't done while I had been on the phone. (Hopefully I would have the chance to catch up without another visit from May.)

There were numerous times when May would just stand and watch me work. She would literally stand over me and stare at me while I went through charts, took phone calls, or handled problems. I had no idea what she was looking for, or if it was nothing but an attempt at intimidation, but I tell you: it is not easy to concentrate when under that much scrutiny. She would say to me "Your better watch it" and "make sure you document; you need to protect your ass." I never knew what she meant when she said that. There were some things that did need documentation from me specifically, but other things that didn't. Issues that directly affected patient care or were disruptive to staff needed documentation. However, transferring a patient to a higher level of care would not require documentation from me. As May would never really explain what she meant, I was left in the dark as to what she felt was document worthy and what wasn't. The other house supervisors never did this to me, which only added to my confusion.

I worked with the flow of the hospital and since I never heard from my boss about complaints, I figured I was doing OK. I tried to tell myself she meant these comments to come across as "friendly advice" but mostly it just felt like a threat. When I reached out to people I worked with at the time, I heard stories about similar things that continued to happen after I was gone. I heard stories about May making charge nurses cry, how she would take nurses aside and rip them apart, or how she would stand at the nurse's station and watch as they worked. What she was looking for, they could never tell. It was the consensus that May wanted to be the center of attention because she was the house supervisor. As if the actual work of nursing, done by people under her wasn't enough, she needed to create emotional crisis. She thrived on fear and chaos.

These incidents were highly stressful. The entire time she was here, the unit was a tense mess. She demanded attention and would do what she could to get it. After one of these incidents would happen to me it would take time to collect myself and get my head back where it needed to be in order to be an effective heath care provider for my nurses and patients. Depending on how bad it was, it could take me quite a while. The other charge nurses reported the same types of things, including needing to take some time to get their heads together after being subjected to one of May's episodes.

After about five years of working at the hospital, there was a big change on the unit. Our medical floor was moved to a new floor of the building, and we combined departments with the surgical floor to become a medical/surgical floor. Before the change, the medical floor had taken surgical patients only if the surgical floor didn't have open beds or appropriate staffing levels. Surgeons, as a rule, like their surgical patients on their own floor, to limit exposure to non-surgical patients. Medical patients tend to have more contagious conditions, and surgical patients are at a higher risk for catching contagious diseases and infections because of their open wounds, recent procedure, anesthesia, and decreased mobility.

We did not have enough physical beds to handle the number of incoming patients. Staffing became an even bigger nightmare. Surgeons left in droves due to lack of safety for surgical patients. About two months later, after the dust had settled, upper management figured out that this was a horrible idea. Surgical was moved off of the medical floor to a different floor. So we shifted back to being medical and oncology. (However, we did still take surgical overflow.) It isn't uncommon for upper management to make drastic changes or decisions without asking the floors or staff. It is also not uncommon for there not to be an office announcement or planning on how the changes will be implemented. Someone (usually not a medical someone) gets an idea, thinks it is a great idea because they had it, and off we go.

After the floors had been split up, and surgical went back to their floor, we ended up keeping some staff from the surgical department. One of the nurses we kept was fairly emotionally unstable. I will call her Allie. With Allie you never knew what kind of nurse you were talking to at any given time. Every shift she would be different, many times within a shift she would emotionally change on a dime. She would swing from one moment where she was team oriented, helpful, and insightful. The next moment she was cussing at the patients and the staff, sometimes under her breath and sometimes right out in the open. She would go from volunteering to help with procedures or admissions to tying up multiple staff members at a time because she was upset about something and wanted them to be too. She was unpredictable, at times explosive, and she was good friends with May; they had worked together for years. So over time, even though I didn't think it could be possible, May became even more aggressive towards me once it became clear that Allie was staying. I didn't put it together right away that Allie was most likely encouraging May, but it didn't really matter. It wasn't something that I could prove; it was just a hunch. So I dealt with it, mostly by not dealing with it.

At some point, May began pulling me aside and saying things like "you need to watch yourself, people are saying things." I would ask her to explain what she was hearing, what "things" were people saying, but she wouldn't elaborate. These occasional warnings from May, in addition to her aggressive and accusatory behavior, only served to put me on edge. I did not know what she was trying to do, or why she would warn me about things she refused to explain or why she would then turn around and act so hostile. After a while, I figured it was meant to keep me off balance. I told myself let it go. I figured if people began saying things that really mattered; my boss would say something to me.

Over time, May became harder and harder to deal with. She would belittle and accuse me of things that were utterly untrue. She would say that I was trying to hold up admissions to the floor, transferring patients inappropriately because of my lack of clinical skill, refusing to take patients, refusing to answer the phone, refusing to be a team player, hiding beds from her. I thought to myself "Hey, this is part of the job. I just need to suck it up." I would occasionally write something to my boss, but May never changed. If I transferred a patient off the unit to a higher level of care, I got a phone call, or she would come to the unit to vent her anger and voice her accusations.

May ranted that I was incompetent, that I was clinically unskilled, and I was a bad nurse, and an ineffective charge nurse. She would question my every decision, and whoever was nearby got to hear the whole thing. If she were questioning me because she had concerns with my logic, or assessment, or require more information, I would have been completely okay with that. But she never really listened to any of the explanations. She would ask redundant questions, bait me, wait until she had found what she thought was a flaw in my logic, then she'd cut me off, and jump all over me, lecturing me that the decision was wrong.

Now to clarify, I don't have an issue being challenged, or having my decisions questioned. This is especially true when I believe that whoever is doing the questioning has my back and wants to be sure we've come to the best solution for the patient. That sort of questioning is a necessary and important part of the house supervisor's job. Because the house supervisor is responsible to the flow of the entire hospital, I expect to be asked questions. After all, the house supervisor knows (or is supposed to know) much more about what's going on with staffing on the other floors, how many patients are waiting for placement, the status of the overall hospital and bed situation better then I do.

At the end of the day the house supervisors have people to answer to as much as I do. The difference was that when I was questioned by the other house supervisors, I felt like it was a collaboration. They asked for information that they needed to have, and they did it without insult or accusation. We were working towards the same goal of maintaining patient and staff safety. When I was questioned by May however, it never felt "clean". It was like handling a wet, dirty, and loaded gun. There was always that undertone of her looking for a reason to go off, trying to find something wrong, waiting to cut me off, jumping at every chance to imply my incompetence, or have some other reason to be angry. There was always a sense of unavoidable explosion.

I acknowledge some decisions need to be scrutinized: we all need to evaluate, learn, and make better decisions in the future. But when every decision I make (even the decision about when I went to the bathroom) was dissected and criticized, it began to overwhelm me. I felt like ALL my decisions when I worked with May were wrong and needed to be under a microscope. I became jumpy, edgy, and second guessed myself constantly. It started bleeding into how I was dealing with other house supervisors. I started second and third guessing even basic decisions, whether May was working that night or not. Even though I continued to get positive feedback from the other house supervisors and my boss, that didn't stop May's influence on me. It was so intense it started to eat away at me. I began calling the house supervisors, including May, to update them several times a night as to what I was doing, the morning staffing, the flow of the floor. Many times the house supervisors were really busy, and my call was not helpful to them, but I still did it. I was trying to get ahead of the feeling that, no matter what I did, it was going to be wrong. And more than that, it was going to cost someone their life.

And then it happened

I wasn't even scheduled to work that night. The floor was short
staffed and I agreed to come in when they called me. It was February,
and cold out. I was having trouble breathing (I get bronchitis almost
every winter) but I had an appointment with my doctor coming up, so
I figured I would be okay to go in that night. It wasn't long into my
shift before I was struggling to breathe. I couldn't talk between my
labored breaths and was nauseous. About four hours into my shift, I
knew I was in real trouble. I called May and I told her I was having
trouble breathing and needed to go to the emergency room. I also
had to tell her that I was having trouble getting someone to replace
me that night. May hung up on me and came up to the floor.

When May arrived, she took me into the break room and sat
down across from me. I couldn't breathe. I tried drinking hot tea
(that normally got me through) but my breathing kept getting worse,
and I was getting scared. May sat across from me and started to tear
me apart. I couldn't believe what I was hearing. In between looks of
disgust at me, she said "there is nothing wrong with you, you are fine.
You just need to calm down. How many days have you missed this
year? They will fire you if you leave." May kept at it. I sat across from
her, trapped in the break room with her, struggling to breathe, and
not having the physical ability to respond. The one or two times I
attempted to speak, May cut me off. She grew angrier as time
passed. She said over and over that I was faking it, I only needed to
calm down, and I was going to get fired. Honestly, I started blacking

out. I literally started losing time. I can't remember everything that she had said. I just focused on trying to breathe. Next thing I know I looked at the clock. It was an hour later. I just sat there in stunned silence.

Between insults and threats, May answered her phone and kept up her house supervisor responsibilities. Then, she would continue her degrading comments about how I was faking it and she was not fooled. I sipped my tea, not really knowing what to do. My condition worsened over the hour, but I just sat there stunned, trying to breathe. I did nothing.

Finally, after an hour of this verbal onslaught, two other nurses came into the break room. They took one look at me and told May I needed to go to the emergency room immediately. I looked like I was going into respiratory distress. May protested, but they insisted. They put me in a wheel chair, and took me down themselves. My wife came and got me. She drove to the closest emergency room. I did not go to the one in my hospital because I was terrified that May would follow. The emergency room took me in immediately. The doctor did a chest X-ray and diagnosed me with severe bronchitis that was borderline for pneumonia. I almost cried. After months of mental and emotional abuse, and an hour of threats and intimidation, I was starting to think that I was faking it; that I did just need to calm down. That maybe I was just having an anxiety attack. But here it was right in front of me, in black and white. This wasn't in my head; this was severe bronchitis, it was nearly pneumonia, and it was real. I walked out of the emergency room so full of relief that it didn't really matter that my shortness of breath had improved only a little. This wasn't in my head. This was real.

After that night

That night in the emergency room, the doctor put me on high doses of oral steroids, antibiotics, three different MDI's (metered dose inhalers, sometimes called puffers) and an inhalant called a nebulizer. I started on five treatments of medication that had to be taken every four hours. It took about two weeks for me to be begin to breathe better and go back to work. By then, though, something had started to happen to me. Something subtle and quiet. Something I hadn't felt coming. I didn't understand it in the beginning. It actually took me a long time to comprehend. On my first night back after my bronchitis diagnosis, I was scheduled to work at six that evening. About two that afternoon I woke up in a panic. My heart was pounding, I had a cold sweat, and felt a sick dread in my stomach. I called off work, confused as to what was happening to me, thinking that it would pass.

This waking in panic began to happen more and more frequently as I started calling out of work more often. The thought of going into work made me shaky, nauseous, I couldn't think clearly, and I could feel my heart as if it were trying to break free from my chest. I wanted to puke, and I wanted to hide. I wanted to be anywhere but the hospital. It took another two weeks of daily call outs before I started to get an inkling as to what was going on. I was in trouble, this wasn't going away, and I needed help. When it began to sink in, I called my boss (I will call her Sue), and set a meeting with her the next night before my shift.

That night I explained to Sue what was going on with me since my emergency room visit. I told her what had happened that night with May, why I hadn't been to work, and what I was currently dealing with. She listened intently. Afterward Sue offered me an opportunity to meet with May's boss. I said yes, but I told Sue that I couldn't do it alone. Sue offered to be there, and I jumped at the chance. The next night, I went into work and met with Sue and May's boss, whom I will call Ruth.

As I started to explain what had happened that night Ruth cut me off, and immediately changed the subject. I was stunned. When Sue tried to bring the conversation back to what happened between May and I, Ruth blamed me for the incident. I found myself defending and justifying the kind of charge nurse I was; explaining how my turn around times for bed placement on incoming patients were some of the best in the hospital. I tried defending my record, letting her know if she didn't believe me, she could check with other people. At that point I just stopped talking altogether. I felt like I was being accused and belittled all over again. My heart sank. I remember walking away from that meeting thinking that May had the total support of her boss, and I was a target of opportunity. I felt that somehow that was entirely my fault. But mostly I walked away feeling unheard, terrified, and a little shell shocked. I felt like that night in the break room had just happened all over again, and now I had no recourse.

A week after that meeting I met with my doctor. I had a lot to talk about. I wasn't sleeping. I was feeling panicky all of the time, even when I wasn't scheduled to go to work. Going to work had become harder and harder. "I need help" I told her. She started me on two medications to get me to sleep and back to work. At that point, I applied for intermittent FMLA to protect my job. I knew that I wouldn't have a job to protect if I didn't get some protection for my calls offs, and fast. The panic attacks kept coming. I would awaken half out of bed, my throat tight, my heart in a rage, repeating out loud "I can't do this. I can't do this." The medications curtailed the attacks somewhat, but not completely. At least the intermittent FMLA helped keep my job safe. But I was stuck trying to figure out what was really going on with me and how I was supposed to stop it. I felt that the whole thing was a personal failure on my part. I told myself that I was being a drama queen. I told myself that I needed to get it together somehow. I felt it was my fault for not being able to pick myself up, suck it up, and get back on track.

When I was able to go back to work, it took everything that I had just to stay functional. On the nights that May wasn't working, I held onto the thought if I could just get through the day, that would prove to myself that I was really okay. I found that when May was working, and came to the unit, I would answer her questions and then pretend I had something to do, giving myself a reason to walk away. About a month into this type of coping, as I walked away, May followed me. Talking to my back she said "I know you are walking away from me." I smiled and kept walking. She kept on saying "My boss told me that you were on some sort of leave because of a talk we had. I know that isn't true." I stopped and turned to look at her. I opened my mouth to speak, but didn't get a word out beyond "Well."

She interrupted me, "I know that isn't true. Look, you are an excellent charge nurse, and I love it when you work…". She kept talking, but I could not tell you what she said. When she interrupted me again, my gut clinched, my mouth went dry, and I started to sweat. When she finished she walked away, and was once again left to pick up the pieces. I felt sucked back into the mess, unable to get away. For the first couple of months after she said that to me, May eased up. But it didn't last. I should have known better then to hope it would.

I tried my best to do my job. I took the medications the doctor prescribed, and I worked to maintain my self-care routines, but nothing seemed to help. The guilt from feeling like a personal failure ate me up inside. What was wrong with me? Why couldn't I just let things go? Why couldn't I get myself together? I was in such a constant state of emotional and mental crisis that I couldn't think of anything beyond what was right in front of me, in that moment.

I found that, as hard as I tried to maintain my focus, I disconnected at work. During my shifts, I found that I would go to the bathroom and hide. I would ignore call lights. I would find excuses not to help the nurses and the techs as much as I had in the past. I relentlessly second-guessed my own decisions and always felt on edge. Situations that I had handled with ease in the past became very difficult. I was embarrassed by my own reactions and behavior. I was terrified that the nurses would find out I was handling things so badly. It became a cycle of guilt and self- doubt, and I struggled constantly.

Right after I went on intermittent FMLA my hours changed. It was a common thing for the group of charge nurses to get together and move our shifts from time to time. We had done it for years, and it always seemed to work out. This time my work week shifted. This change wasn't planned, or premeditated, it just kind of happened. I liked the change, as it took me off of Saturday nights. The only thing that was new on the schedule was that one night a week I was working with a different secretary. I will call her Nydia.

Unit secretaries are central to the smooth operation of any unit. I depended on my unit secretary as much as my right arm, if not more. When I was in an emergency situation, or a code, it was the secretary who was calling everyone in. When a code happened, it was called into a central dispatch system, which then pages it on the overhead intercom so that the whole hospital hears it. Most times it was the secretary that made the initial call, followed by calling the doctor, calling respiratory therapists, getting the paper work together, and calling the bed board to let them know what higher level of bed we're going to need. When I was running crazy with my end of things, it was the secretary who called housekeeping, the emergency room, the bed board, and any number of other departments. They were (and are) the life blood of the hospital in so many ways. I can't say enough about the men and women who are unit secretaries. When it works well, it is beautiful, seamless, and completely patient focused.

But with Nydia and I, it never really worked well. We had worked together off and on over the years, but we did not get along.

To me, our personality differences had always been secondary to the question of "could we work together effectively?" In the past, Nydia would leave the unit for lunch without telling me, regardless of what was going on. Once she left in the middle of a code (a patient was actively dying and we were trying to make them not dead). It wasn't that Nydia didn't know that a code had happened, because I had asked her to call it overhead. As I was coming out of the patient's room to get the intubation kit, Nydia was walking onto the elevator to leave. She would leave during react team situations, where we would need to get a patient off the floor immediately. She would disappear when I was doing my rounds. She would walk off and leave the desk unattended with no one to answer phones, respond to call lights, or assist the doctors.

It wasn't that I cared that she was going to lunch. What I objected to was that she would never tell me when she was going to go. This left me with no time to arrange to staff the desk in her absence. Originally we worked together so rarely I just let it go; it wasn't worth the fight. I would occasionally notify Sue if her behavior started to affect patient care but, for the most part I told myself "just get through this one shift." And I let it go. After my shift changed, we were working together once a week. On its own, this was a problem, but not a big one; this could easily be solved. But now that I was fighting to make it through every shift, it was hell.

On top of leaving the desk (or the floor) at random times, Nydia would either ignore my questions or intentionally speak so quietly that I couldn't hear her. At first, I thought maybe this was personal. I knew that we did not get along. So I asked the other charge nurses what they were experiencing, and they all said she did it to them as well. One charge nurse explained that if she really needed the information, she would walk over to Nydia, stand over her, and ask again, louder. There was NO WAY that I was going to do that. I felt if I stood over her, it would only make a bad situation worse. As we started working together more consistently, I started having to write emails to Sue. I explained that I could not hear Nydia, that she was ignoring me, and that she would walk out when it was inappropriate and unsafe to do so. I reiterated that I didn't care when she went to lunch. (In fact I was a strong advocate of the necessary break that lunch provides.) I just needed to know when she was going so I could arrange for coverage or respond to issues accordingly. But the mumbling and not responding to me when I addressed her directly, that was a bigger problem that directly impacted my job and patient care.

I wasn't trying to make conversation about her sweater, the weather, or what she had cooked for dinner the night before. I was asking about the patients, their care, their safety, if a doctor had or had not called back, etc. This was important information that I needed, and I didn't have time for what I considered petty games. I hated to admit it, but it was starting to get to me. I would ask a question, she wouldn't answer, and I would hang my head and sigh. I never confronted Nydia on this. I didn't accuse Nydia or berate her. I just got frustrated. I believe that if I showed that I was angry it would only serve to make a tense situation worse. I would leave the desk and walk around the unit, trying to figure out how to deal with her. I didn't know what to do. I kept writing my boss Sue, and hoping for a change. I felt if I contacted Sue, surely some change would happen. But in those actual moments of having to work with Nydia, it was worrisome because I truly felt that my job was compromised, and because of that, the patients and nurses that depended on us were as well.

I think that it is relevant to state that I am a 5 foot 9 woman of size, and I am not exactly overly feminine. My appearance alone can be off-putting and intimidating for some people. Getting angry or coming across angry just tends to make things worse. So I grappled with how to communicate the importance of the information that I needed without seeming intimidating or angry. It is not that I want to be approachable because I want people to like me, or think that I am a nice person. What people personally feel about me on the job wasn't as important to me as working together to get that job done. My social life is outside of the job, so my self-esteem wasn't dependent on the job as it can be with others.

However, I did, and still DO CARE if others' thoughts and preconceived notions about me get in the way of us doing our jobs, or if it impacts patient care. I did not ask Nydia to like me. That didn't matter. I did need her to communicate with me in a way that was safe and effective. That absolutely mattered. What I cared about was knowing if I needed coverage for her position or was working without backup. If I was doing chest compressions on a dying patient, knowing that I was alone was kind of important. Even after repeated contact with Sue about the issues, there was no change in Nydia's behavior, and I was getting really frustrated.

Before long, I started hearing rumors, not the kind of silly rumors that are funny stories that build a sense of camaraderie, but the sort of destructive team splitting rumors that can wreck morale and affect team work. The rumors I heard included that I was telling Nydia that she wasn't allowed to go to lunch. And that I would send her home early at the end of the night to intentionally short her hours. Neither of these were anywhere near true. Sure, I wanted to know when she was going to lunch, and if the end of the night was slow and we were caught up, she would ask to leave early. And why wouldn't I let her go if we were slow? But I never initiated the conversation and I never asked her to leave. It was always her choice. These rumors seemed so ludicrous that I gave them no credence.

I am a huge advocate for all staff going to lunch and taking their breaks. I always have been. If medical people aren't careful about breaks they can easily go eight or ten hours without stopping to eat, drink, or go to the bathroom. Not taking breaks slows down the mental processes, makes people less able to handle complex situations, and, most of all, it makes people grumpy. I always pushed staff to eat, drink, and take breaks. The idea that I would stop someone from going to lunch was ludicrous to me, literally just beyond me. As far as shorting her hours, I shook that off as well. In my opinion this was a "she said-she said" situation. I felt that nothing I could do or say would really matter. After thinking about it for a while however, I wondered "what if people really believe this?" I decided it needed to be taken to Sue so that she could advise me. I remember saying to her "this is so ridiculous I can't believe that I am coming to you. But if she asks to leave at the end of a slow night, what do I say? I mean, do I keep her for no reason other than rumors, or do I let her go?" Sue said to let her go, but make sure to email her the time that Nydia left and that she had requested to leave. I said "Okay, fair enough." Stupid, petty, and silly, but I would do my part.

It was about two weeks later that Sue asked to speak to me again. She said that HR was notified about something and that they would be calling me. I said "okay can you give me a clues about what is up?" Sue said that she couldn't. I got the call from HR. They asked some vague questions at first, but then as I mentioned Nydia, they narrowed their focus. It was obvious that Nydia had reported me about the lunch hour, being shorted hours, and whatever else she felt was relevant. More than that, it was implied to me that I was always angry when I was dealing with Nydia. This made her feel scared and created a hostile work environment for her. HR asked me if I got angry on the job and I thought about it before answering. I remember thinking "well everyone does". So I said "yes." Then they asked me "what do you do when you get angry?" I said "I just focus on work, or take a walk around the unit, but it depends on the situation. If I were frustrated with someone, and depending on the situation, I would discuss it with them in private." Throughout my conversation with HR, they kept saying something along the lines of "don't get angry at work." So I said "okay."

I mean what was I supposed to say? Because they were not allowed to give me names or even tell me about a specific "thing" in my behavior to correct, it was all vague. I was left more confused than ever. I didn't really comprehend what had happened. I could just piece together what I thought it might be. But the part that made me really scared was that idea that I was coming off angry at work. I knew I was frustrated over the Nydia situation and I felt that I was not handling it very well. Now it was confirmed that I was not handing my feelings as well as I had thought. When I was working with both Nydia and May, I was really struggling. But HR gave me no tips, advice, or help in how to deal with the situation. I told them I had spoken to other charge nurses, and how they were handling the situation, but that I was not comfortable confronting Nydia in that way. Their answer was "don't get angry." Great. "Don't get angry." Got it.

Even after that call with HR, the gossip about me cutting Nydia's hours and refusing to allow her to go to lunch didn't stop. She did occasionally use a voice that I could actually hear, but it wasn't consistent. She would slip right back into her old behaviors, frequently. Then she stopped going to lunch altogether, which didn't make any sense to me, especially when there was plenty of time for her to go, and she had complained about it. I figured it was her choice. I had to sign her lunch time sheet when she didn't go. Then she started going to lunch sporadically, but my request to have a timeframe at the beginning of the shift was ignored, and she would just get up and go. I kept writing to Sue, but nothing changed.

What was changing was that I was becoming less and less capable of maintaining my composure at work. May continued her bullying. Nydia kept mumbling and abandoning the desk without warning, and I started to crumble. My ability to deal with complex and highly stressful situations deteriorated as I fought to hold it together.

The ripple effect

After the night with May, and the ongoing issues with Nydia, I wasn't the same. When I was able to go to work, I struggled deeply. I kept going to my doctor, who was wonderful and helpful. She prescribed medication that I was completely compliant with taking, and she kept my intermittent FMLA without any problem. But that still wasn't enough. The thought of going into work would wake me up from a dead sleep, jittery and frightened. I would pace in circles in my bedroom saying "I can't do this. I can't do this." I would have nausea and diarrhea. I was terrified that if I went into work in that condition, something would happen; that I would not be able to get it together, and someone would die. I couldn't risk that, so I avoided going to work as much as I could. I strove just to have a day at home or at work where I felt safe.

Calling out of work so often devastated us financially. I am the breadwinner in our household, and as we fell further and further behind in our bills, the guilt and self-doubt flourished. I found myself in another vicious cycle. We were behind in everything. Bill collectors were calling all of the time. We were constantly on the verge of having all of the utilities shut off, the car repossessed, and the house foreclosed. We couldn't afford propane for heat or consistent food. It wasn't because I didn't make a good wage. As a nurse I made good money. But I couldn't pull myself together enough to get through the doors of the hospital in order to earn that wage. The FMLA paid out of my emergency fund hours, which I had over 300, but at a lower rate. So although my job was mostly protected, my checks got smaller and smaller. I berated myself for being a coward and a failure. I told myself I should have felt lucky and kissed the hospital's feet for the kind of money that I was making, and the opportunities they were providing. What was wrong with me? It was tearing at my very core. I felt completely useless. Every week I would tell myself "okay, this week I am going to turn this around. This week I am going to get it together, this week is the end of this nonsense." Every week I would fail again. Sometimes it would start a few hours before I was supposed to go to work. Sometimes, it would start the night before. I would feel the flutter in my chest, my pulse began to race, and my brain would flood with fear. I would take something for anxiety, just to get me to sleep. I thought that if I could sleep through it, I could "trick" myself into going to work. Sometimes it worked and sometimes it didn't. As time went on it stopped working completely.

There were times when I would lay in bed completely frozen. My mind would race. I started to think of ways to kill myself. I had a barn and rope. I had enough anxiety medications. I had means and I had a plan. I remember one day in particular as I lay motionless my wife came in with the phone. She insisted that I take it. I did and put it under the covers with me. It was my chosen family, Sam. I quietly told him that I was having "thoughts". Being a therapist, he knew what that was. He made me promise that if I ever was going to follow through with it that I would call him first. I promised, not truly knowing if I could. What made me not go through with it was the thought that my love, my wife would be the one to find me. The thought that no one would be there to take care of her. So, although the thoughts didn't go away completely, they did lessen some.

When I first went back after the incidents with May and later with Nydia, I would have panic attacks in the middle of my shift at night. If there was any kind of infighting, or I had to talk to someone about his or her behavior, I would manage to get through it but inevitably fall apart and need to go home afterwards. It became a fight to get through the doors at the beginning of my shift and a battle to stay the whole night.

I would burst into to tears at things I used to handle with ease. I would go to the bathroom, sit on the toilet and rock myself. I would hide in the break room staring into space. Then it happened, the thing I had been dreading. I was working with another nurse on something that she needed. It wasn't that complex, if I remember correctly, and I suddenly felt deeply overwhelmed. I started to rock myself back and forth in my chair right in front of her; trying to wrap my brain around her very simple request. She stopped what she was doing, took a step back, and stared at me. The she said "ooooo, never mind, you don't have to rock yourself." She saw it and others did to. I was mortified. My secret was out, and I felt even more exposed.

It was a few days later when I had a full blown panic attack, right at the desk. I will never forget it; it was five o'clock in the morning. I had to get up and prepare the assignments for the oncoming shift. I couldn't move. I literally could not move. Panic flooded my brain. I couldn't do it. I couldn't do the assignments, the simple shift assignments. I started to think, trying to clear the storm. "Okay, what's the worst that would happen if I don't do this?" I could wait until the oncoming charge nurse came in and let them take over. That was the worst that would happen. I remained sitting, fear flowing out of my pores. My heart felt like it was going to explode right out of my chest and land on the desk before me; my brain was unable to organize itself, it was on fire and frozen at the same time, while my legs were detached, heavy, planted and unmoving. It was terrifying. I had to calm myself.

When I could get a thought together I told myself, "you are having a panic attack right now, you are okay, this is just a physiological response to your fear. On the count of three, you are going to stand up and just do one thing, one thing only. Just erase the current assignment that is all, just one thing." On the count of three I stood up, which surprised me. I hadn't been sure that I could. I walked over to complete the shift assignments and, before I knew it, my shift was over, the day charge had arrived and I was leaving out the elevator. As far as could tell, no one knew; no one but me.

When my wife picked me up from work, I sat in the passenger seat and started out the window. Nothing had ever frightened me so badly. At that moment had there been a code, or a react team situation, I have no idea what I would have been able to do. I would like to say my ten years of dealing with emergency situations and training would have kicked in. But I honestly don't know. I couldn't tell you what would have happened. Had May decided to come to the floor, or call and start in on me, I have no idea how I would have reacted. I had been scared before, but now I was deeply terrified. Things went from bad to worse. I ended up on five different medications to help control my anxiety, my now full-fledged depression, and the PTSD that was taking over my life. I developed GERD, and for added fun I had diarrhea every night that I worked.

As crazy as it sounds now, at no time did it occur to me to go to HR. I thought that if it was serious enough, surely Sue would have told me to go, or would have made sure that HR was present when we spoke to Ruth, May's boss. As I look back on it now, that was probably a very short sighted choice. I just did not see going to HR as something that would have been helpful. The biggest reason I didn't go was because I was afraid that I really was blowing things out of proportion, and after talking with them about Nydia, their ability to truly help seemed minuscule. What where they going to tell me? Don't get angry?

When everything started to snowball, after the night with May, I was resistant to go to counseling. I am not opposed to counseling or therapy. I wholeheartedly advocate it. But in this case, it seemed to me that this was a personal and moral failing on my part. This was just me needing to "get it together". This wasn't about what someone else was doing. My pain and my problem was because I couldn't hack it. I was convinced that if I could just get it together then everything would be okay. I realize now that I was terrified that the counselor would tell me that it really was my fault. Not to mention as we were sinking financially, the thought of attempting to afford another co pay was overwhelming.

Even as I was struggling with feeling like a failure, I couldn't fathom how the low performers on the unit never seemed to be bothered by anything. I had three nurses who worked with me for years, and while I am sure that they were good human beings, I wouldn't want them as my nurse if I were hospitalized. I knew on nights when I worked with all three of them, at least one of their patients was going to code or need the react team. I knew I could ask any of them a question at any time about any of their patents, and they would not be able to answer me. I am not referring to in–depth stuff here; I mean basics like code status, diagnosis, and age. They wouldn't know and what's more they wouldn't CARE that they didn't know. They seemed perfectly comfortable with this level of detachment and indifference. So when I compared my perception of my own performance with theirs, I was only more confused.

I am not saying that I was super nurse, but I am saying I was a higher level performer. Without a doubt, I was more skilled than many of them in both procedural skills and assessments. I was able to answer more questions about their patients than they were, and I was in charge of forty to fifty patients at any given time. In a code or react team situation involving their patients, I would take the lead because the three of them would either fall apart or disappear altogether.

With those three nurses, I had to make sure to catch them and escort them back to their patient's room if there was an emergency situation, or I would not be able to find them. As much as it is hard for the medical community to admit, nurses come in ALL skill levels. Skill level has NOTHING to do with their years of experience or their level of education. It has to do with that individual nurse's drive to be a thorough, thoughtful, and engaged person. I have worked with LPN's that were more skilled than RN's with bachelor degrees. I have worked with new nurses that were better at assessing patients and utilizing their skills than nurses who have been on the job for ten years. It happens, and it is more common than you think.

Some of what I felt about them was no doubt, grounded in my own ego, but that didn't make it less true. And it made things worse for me. While I was falling apart, they seemed perfectly fine while they half did their jobs, just as content as anything. I kept thinking "what is wrong with me? Why can't I just go along like they do, and not engage the way they do? How is it that these low performers are able to handle the job, day in and day out, AND work overtime, but I can't get my shit together?" In many ways, I felt their ability to come to work (not their preference on the job, just their ability to show up) out performed me a million times over. Though they would rarely finish their work, and what they did finish was poorly done, incomplete, and in many cases needed to be redone, at least they SHOWED UP. It was more then what I could say for myself.

All of this led to the deterioration of the pride I used to feel in my work ethic and job ability. I became more and more nonfunctional. The first year after the incidents with May and the start of Nydia, I called out more often than I went to work. Not only did I max out my FMLA in ten months, but it also created a very strained relationship with my boss, Sue. She was trying to be supportive; after all, after my FMLA was maxed out she could have fired me. But Sue kept trying to work with me; she kept me on, to see if I could pull myself together and get back to the job at hand. Sue had every right to (as well as cause to) fire me, but she never did.

Finally, after that first year, I started going to therapy. I started going once a week, then twice a week. This included going to therapy on the morning of the nights I would have to work. I was given techniques to help calm me, like physical methods of self-soothing, calming thoughts, and thinking of safe, happy places. These techniques helped. They gave me something to "do" when I felt the panic coming, or when I felt like I was going to lose my mind. My therapist told me that if I feel a panic attack coming, go walk a flight of stairs. Apparently you can't panic and exercise the same time. It didn't make everything go away for me, but at least I had a plan to gain some control over myself. The goals I was working on with the therapist were simple: go back to work, come off of FMLA, and eventually come off at least three of the five drugs I was using to maintain my tenuous functional level.

This was not to be.

Over time the situation with May became worse. As the emergency room was pressured to get patients out, they were coming to the floor in droves, and hardly ever appropriately. On a nightly basis, we started transferring three or four patients off of our medical/surgical/oncology floor to higher level of care. One night in particular, we transferred three patients off of the floor back to back. One was set for transfer before I arrived. They had high blood pressure. By the time I came on for my shift the patient's blood pressure had returned to normal, so both myself the patient's primary nurse spoke with the doctor on two separate phone calls asking for the doctor to leave the patient on the floor. The doctor insisted on the transfer. The second patient was having all of the vital sign changes and behaviors of going through a withdrawal. They had a high heart rate, high blood pressure, sweats, tremors, and agitation. Even through their blood alcohol level was zero, the doctor agreed to transfer to a higher level of care. The third patient went into respiratory distress and react was called. The patient was put on a bi-pap to facilitate breathing, and the react team talked to the doctor.

That is when I got the call from May. She degraded me over the phone. I tried to explain what was going on, and she cut me off and hung up on me. It was only minutes before May came barreling down to the floor. I could feel her before I could see her. I was sitting at the desk going through the carts, getting things ready for the react team to move the patient to a higher level of care. She started accusing me of not knowing what I was doing with the transfers, how these transfers were inappropriate and how from then on out she would have to know about every transfer before they were moved. It was then that the head of the react team walked the corner and came into the nurse's station. May stopped. I just looked up at her and said "Okay." And went right back to what I was doing. May went over to the chart and spoke to the react person. She pushed to get the patient to stay, but the need for bi- pap was an automatic transfer. She collaborated with him.

The react team continued to tend to the patient until they were transferred off of the floor. As May was leaving, she warned me "you better cover your ass. People are going to get you in trouble for all of these transfers." I ignored her, surprised at my own reaction. Turns out it was the first time and the last time I would do that.

The next day

The next day I came into work, I was on the floor for about ten minutes when my boss, Sue, came. She asked me to come with her. I left my bag and we walked down to HR together. As we rode down I was grateful. I thought "I have missed so many days, I am glad that the security guards had not been called to escort me out."

Once in HR I sat at a table with Sue in front of me, an HR person to my left and one at the far end of the table to the right. The HR person sitting next to me put a piece of paper in front of me. I read it, couldn't understand it, and read it again.

The first offense was the three transfers from the night before. I slowly tried to find my voice and explain. Sue cut me off and said, "you have transferred inappropriately before." I just looked at her, confused. "Remember Thomas?" Sue asked. I did in fact remember Thomas. It was three or four years ago; he was an oncology patient. I had no oncology training at that time and his labs (blood work) were dangerously out of control. I learned from that and I had not transferred an oncology patient due to lab work since.

The second offence read: reading at the desk. I looked at Sue. I had studied during my break time as I was in school, but as soon as Sue had told me not to I stopped. I didn't bring in books, or even magazines.

The third stunned me completely silent. It read sexual harassment. I couldn't take the words in all of the way. I couldn't believe what I was reading. I was an out lesbian. I always had been, but besides answering direct questions, I never talked about it. I wouldn't lie, but I would not instigate either. I tried to formulate the words around what I was seeing. The HR lady said harshly "I am married to a man. I never share what we do in the bedroom." I looked up at her confused. My mind was running crazy. What? What was she saying? What? I was too stunned to respond. I had no breath, no words. I fought to remember anything that I could have done that would have been perceived as sexual harassment. I came up completely blank. I started to sweat. This wasn't me, they had the wrong person. My throat tightened in fear.

The fourth line read "demoted." Demoted? That made no sense. I sat back, I remember saying "this is extremely serious." The HR lady said "yes."

But, if I had done all of these things, I had received no warnings about them, and let me be clear here – not one single warning-- if they were so serious as to cause a trip to HR, why not just fire me? Why not talk to me about my behavior when they had called me about Nydia all of those months ago? And why keep me on the floor instead of suspending me? If I was dangerous enough to be classified as someone who would sexually harass others, why ask me to go back to work that night?

I signed the paper. I didn't know what else to do. I got up, called my wife to come pick me up, thanked the HR people and my boss, and left. The next day I put in my two weeks.

Since then I have talked with other nurses about these charges. I don't understand them any more now than I did then. I had worked with my crew for eight years, and with some of them for another two years before that. I had never had a complaint about inappropriate behavior before. And HR, being HR, was not able to tell me anything. I thought long and hard before I wrote this last part. I wasn't going to. I was going to white wash over my trip to HR and the outcome there. But then I thought, no, this is what happened.

If I were you dear reader, I don't know what I would be thinking at this moment. Maybe you would think that I am incompetent, or that I am a pervy skeeve. I don't know.

What I do know is that I didn't transfer three people inappropriately. It was by the book, every one, down the line. I hadn't brought in a single piece of reading material for myself in seven years. And in all my years I have never sexually harassed anyone. What I know is that they wanted and needed me out. I don't blame them for that. I should have been fired when I could no longer do my job, and my call offs were more than rightful cause. So why the other complaints? I do not know. Maybe to scare me or maybe to get me to walk. Well, it worked. I put in my two weeks the next day.

Then something else happened. I got free.

Starting over

I have no regrets. My wife can't stop smiling. There hasn't been one day where I thought to myself, "I should have stayed, gutted it out." I admit we were lucky when it came to the logistics of my departure. When I left I drained my 401k to give us some breathing room. You might caution me about my financial recklessness, but I tell you having that moment to breathe was a blessing. It felt like my first breath in years.

Over the next two months I was able to let whatever feelings and emotions I was having just come. At first I glossed over my feelings by keeping myself extremely busy. Later I was able to calm down and just let myself feel. I expected to feel like a failure, worthless, like I didn't matter, and like my nursing didn't matter. And I had some of those moments. But other things came as well. Things like the huge sense of relief, like the world had been lifted from my shoulders. I felt like I could breathe again. People who knew what I did for a living but had no idea the amount of pressure I was under told me I looked wonderful. Those same people told me they had actually been worried about me for a long time, and that leaving the hospital was the right thing. I loved hearing that. The support was tremendous even though they had no idea what had happened.

I began looking at all of the things that working at the hospital had made harder. I had stopped going outside. I had two horses that I hadn't worked with in over three years. They had become my pasture pets, and I loved them for that. But while I was working at the hospital, I didn't have the emotional ability to be with them. I couldn't fail them too. Over the years they had remained fed, watered, and safe. Not with as much food as I would have preferred at times, but enough. I hadn't walked in the pasture, or cleaned my saddle in years. I hadn't gone for a walk that didn't involve feeding the horses and then going back inside the house in well over eighteen months. I had stopped doing the things I loved because I couldn't face those things, because I couldn't face me.

So, very, very slowly I took those things up again. And I do mean slowly. First I cleaned my tack room, just swept the room. So to understand the magnitude of this, let me explain. I have identified as a horse person since I was 18. When I was finally able to buy a house and have my horses on my property, it was like going from being a kid to a full-fledged adult. It was amazing for me to be able to sit out on the porch and watch the horses in their morning routines. Over time, when I worked in the hospital, as I began to erode, those moments of watching the horses meant that I was still a person. As I fought to go to work and began to make less money, it means that I couldn't afford to buy as much hay, and forget getting grain. The guilt over not being able to care for my two horses in the way that I felt they deserved became overwhelming. I started avoiding even going into the tack room because it only served to remind me of how I couldn't get myself together. So that day I went back into the tack room and swept. It was more than just a broom on the floor. It was about trying to remember that I was a person again, that I had worth.

In the beginning it was only as few sweeps before the act itself became too overwhelming and I had to head back to the house. So even those halting moments of sweeping the tack room was a coming back to me. I had been so overwhelmed for so long by what was going on at work that I did not have the physical or emotional energy to do anything on my off hours except sleep. And now here I was, broom in hand, inch by inch, clearing my space. It was transformative, and truly started my healing. It was the valley between what I had been and what I was becoming. After that, I walked all the way from my front door to the road. A whole hundred feet. Then I would get too scared and have to come back inside. It was as though the wind was too overwhelming on my skin. But it was progress and I held on to that.

It was when I started looking for another job that the feelings of failure became intensified again. I was terrified of having a supervisor like May again. In one interview they asked me "what would you do if you and your bosses didn't get along?" It was a good and valid question, but I panicked. I almost burst into tears. I remember clasping and rubbing my hands together, lowering my head, and saying "I would focus on the job; I would put everything into that. But sometimes those things just aren't fixable." The rest of the interview I was unfocused, sweaty, nervous, stuttering, and lost. It was horrible, and I wasn't ready. Luckily I was offered a job as a clinical educator and I jumped on it. It was a lot less money, but I was okay with that. I was barely making enough to survive on anyway, with all of my call offs. The second week I was working I got kidney stones. I tried to come to work, but the pain was unbearable. Two months later, they let me go. It only reinforced that I was not a good person. That I couldn't keep a job. The next month, I got a job in hospice.

I loved hospice. I loved the patients, the travel, the work. It was so refreshing to have time to spend with people, to talk with them, to truly get them what they needed. Five months in I got a new boss. She called me up one day and started yelling at me. She was manipulative and abusive, she screamed that she was going to fire me, cut my hours, that I didn't care, and I was a bad nurse. Her threats of "what if that was your mother?" came through clenched teeth. I tried to explain that with my full docket I wasn't able to stop what I was doing and go see a new patient three hours away. But I would be happy to see her tomorrow. I reminded her that when I took that particular patient load I had informed her of my limitations, since I saw the out of reach patients in small towns and in the mountains. It only incensed her. We had talked about this, I had voiced my concerns, she had agreed to my limitations. That didn't matter, she kept screaming. After about 15 minutes I blacked out. I remember the call ending and when I put down the phone my hands were shaking. I had a job interview the next day, and a new job within the month.

I have been here ever since. I have a great job, a great team. The past came back to me when I first started the new job. Anytime I had to talk to my boss I found myself stuttering, shaking, and terrified. It took about 5 months for that to stop. My boss never yells here. When I came onto this job I became afraid that I would repeat old patterns that I had in the hospital, that I would call off and fall apart. That it really was just me. That didn't happen. I come in every day, on time. I do my work and when I need time off, I get it in advance. I make much more at the new job than in the hospital. And I look forward to coming into work every day. I have regained that sense of pride that I had in being a good nurse and a good employee. It feels so good, so empowering to know that I can do this; that when I am treated well, I can work hard and thoughtfully. This job has been a gift.

I have worked through some things, and I am still working through other things. I have significantly reduced the medications I require to get through the day. I am down to one. I know that there is a part of me that may never fully heal. I may always have a seed of doubt and fear of failure. The question that I get asked a lot is why haven't I gone to a lawyer? It is a legitimate question. There are two reasons why I haven't gone. One, I can't relive that again. With lawyers come depositions, statements, questions, hearings, and cross examinations. Just thinking about it makes me break out into a sweat. The other reason is that I still live with a sense of guilt that this was somehow my fault. I am afraid that when the lawyers get all of their information, they will laugh me out of the courtroom for being weak and stupid.

I know there are people out there who are going to say that I don't exist, or that this story was greatly embellished, or even an outright lie. They are going to say that I was too sensitive, or that I took things the wrong way. I know this because had I read a similar story from someone else years ago, I would probably have said those same things. But for those of you who have read this and are living through the same kind of nightmare, know this: I am real. Everything that I have written here did happen. Did I fail on some things? Sure, I was never perfect. But I put my heart into what I did and onto the people that I worked with. I believed in them when they couldn't believe in themselves. I tried with all that I was to give them as good of a night as I could. I am still in contact with a few of them, and I hear that nothing has changed. May is terrorizing everyone, except

Allie. May still leaves people in tears, still rips floor nurses and charge nurses to shreds, and still enjoys spreading rumors. Nydia is still walking out whenever she wants to, and low talking to the point of not being heard. In essence, my being there and my leaving didn't really matter to anyone but me.

To you, the other nurses out there who are slowly losing your minds because of what is going on around you, take heart--- you are not alone. What is happening to you isn't your fault. You have the right to stand up and say "no". If you aren't able to do that, just know that somewhere out there is someone who understands what you are going through; someone who truly and thoroughly understands the depth of what bullying and hostility can do to your sense of self and your sense of safety. You are not crazy. You are not weak. You are not less than anyone or anything. No one should have to go through this. It has been going on long before you, and will continue long after you are gone. Please know that it really has nothing to do with you. Their bad behavior says more about them than it ever will about you. When you can, let your voice be heard. Talk, write, start in a whisper if you have to and let it grow into a roar.

Most of all never forget—YOU ARE NOT ALONE.

The caring profession

      I used this writing as my voice. It is my way of saying that what happened to me was wrong, what is happening to others is wrong. It is my way of saying to the bosses out there who encourage or ignore situations like mine, "what you are doing is more than wrong, it is criminal." I can hear what some of you are thinking, "this can't be; nursing is the caring profession. Nurses are nurturers and SURELY they nurture each other. Sure, we have all heard about how nurses eat their young, but that is an urban myth, like alligators in the sewer or the hook man." If you believe this is a myth, then you would be right.

Because bad nurses don't just eat their young, they eat everybody.

      There are however, reasons why this behavior is not only tolerated, but actively encouraged in the nursing profession. First, it is only recently that we have begun to look at the issue of workplace violence. The terms "lateral violence" (violence against someone who is considered at the same level as you) and bullying have come into nursing like a fad, complete with yearly trainings. However, it is very difficult to find anyone actually dealing with these issues in hospitals. How can you really police or even document behavior when so much is inferred or implied? What is the difference between an actual bullying situation and someone who is having a bad day? How can you separate someone who is being victimized from someone who insists on seeing the world though his or her own negative filter? There is only so much that a workplace can do to actually regulate employee behavior.

If a company really encouraged people to report bullying, and then fired or suspended everyone accused of bullying, there would be no employees. First of all, all of us have bad days. All of us have co-workers that we don't get along with, and all of us have times where we come across differently than how we intended. Secondly, if nurses are busy infighting and focusing on tearing each other apart, they can't effect change in the system as a whole. You can't come across as an organized front if you can't agree with each other. So the bigger, more dangerous issues go unchallenged; issues like staff to patient ratios, lack of supplies, and increase in workload that negatively impact patient care. It is in management's best interest to keep people bickering amongst themselves; that way they don't have to implement real change or be held accountable.

Third, it increases the overall sense of fear and tension. When staff are focused on doing whatever they can to keep their jobs, they stay quiet and compliant. This is different than keeping people focused on each other; it is about utilizing small areas of control to affect bigger areas of control. For example, nurses and techs are now required to carry phones with them at all times, and they are pressured to answer them even if that are on a break or at lunch. So even though they are technically off the clock, they are required to still be available. This leads to nurses and techs taking five-minute lunch breaks or bathroom breaks, but not putting in their time for "no lunch" because management says that they have to take their breaks no matter what. This, in essence, forces employees to lie and work off the clock; yet another offense for which they might be fired. So they stay quiet. So the question becomes, "if I could get in trouble for taking a potty break, why would I risk talking about hospital policy?"

Lastly, it weeds out the problem children. By problem children, I mean those who the manager or administration do not like. It is an easy way to get people to leave. If the problem child is made to be the object of the bully's attention and then management encouraged or ignored the effects of the bullying, that is a simple solution to a complex problem. That person may be impossible to fire based on their performance, but a bully can make someone's life so miserable that they quit. This saves the company time, effort, and money. The company doesn't have to pay out unemployment and management can bring in someone that they consider to be less troublesome or cheaper.

What I have experienced is that for some people, bullying others and having an unstable and hostile work environment feels good. They want their people afraid; they want their people jumping every time the phone rings, or the bosses hit the floor. They enjoy the sense of power it gives them to be able to degrade others. It makes them feel superior and gives them a rush. Emotionally eating people up fulfills them. To these people, their actions aren't about bullying or intimidation at all; it is just how they get things done. To people like this, others are just asking for their anger because they are viewed as failing and they deserve it.

How many times have you heard a person like this brag "I really let them have it", "I really had to show her who is boss" or "I had to show him that I was the one that makes decisions." There is a feeling of pride that comes along with this for them. Over time people like this tend to escalate because the day to day bullying does not give them the emotional sense of fulfillment that they need. So they will take it up a notch here or there, to get that feeling of control back.

Before you know it, you have a boss that goes from threatening people individually to yelling in the middle of the nurse's station, and a whole slew of people following in their footsteps.

If you choose to stay and are looking for a place to start fighting the harassment, please understand that this can be tedious, emotional, and time consuming. It also, regrettably, may still end in no change. The only real way to actively deal with bullying is through consistent, reported, detailed, and time relevant journaling. This is a journal that should be given both to your boss and HR at the same time. Remember to keep a copy for yourself. Use only the facts, keep emotion out of it. Something like this:

01/17/2--- 12:54 received a phone call from JS. JS asked about staffing and beds. When I attempted to explain our floor situation, she cut me off and would not let me continue. Throughout the conversation, as I continued to attempt to explain our staffing and bed status, JS escalated her tone in an accusatory manner stating "-----". I calmly attempted to answer her questions, but she hung up the phone as I was talking.

Notice that I only wrote exactly what she said, noted her tone, noted when she escalated, and used exact quotes. The escalation is given more merit as she hung up the phone as I was talking. I stayed connected to the facts about what she was saying, how she was saying it, and her behavior. In other words, I did not justify my position, I did not interpret her words, or attempt to inject motivations. I didn't write "she hung up angry."

I know that it can be painful and demoralizing to relive these moments every day, but it is important for you to write them down while your memory is as fresh as possible. Any discrepancies in your record can cast doubt on your version of events. The reason I suggest you giving the journal to your boss and HR at the same time is for political reasons. There are always behind the scenes politics going on. Even if you are positive they are not, there are. Although it is important to keep your boss informed, it is just as important that HR is involved for several reasons:

1) Most likely you are not the only one being bullied.
2) HR is directly responsible for dealing with bullying.
3) HR is required to track and investigate behavior.
4) If your boss overlooks the incidents, doesn't want to cause waves in other departments, or "accidentally" loses the paperwork, HR can be the "bad guy" and your boss doesn't have to be a part of a confrontation.

Now here is where the reality gets ugly. Everything is about perception. It can be difficult sometimes to identify whether or not you are being bullied. It can be much harder to demonstrate that to others. The perception of both parties as individuals, the corporate culture of where you are working, and even different sensitivities, can affect the perception of what is being conveyed. The more precisely you can document the actions, the more consistently you can show a pattern, the more effectively you journal, the more reliable your side of the story is. Something else to keep in mind is to turn the journals in as soon as you establish a pattern. If you wait too long, it could be questioned as to why you did so.

Staying is only one option though. If you choose to leave, please know that it doesn't mean that you are weak. It just means that you need to be better to yourself. When I worked at the hospital, I truly felt that staying was the only option that I had; that I couldn't afford to leave, and there wasn't going to be anything out there for me if I left. When I look back at it now I can understand why I felt that way. And I was thrilled that I was so wrong.

Afterward

It has been a year since I left the hospital. Since then I have found an incredible job that I love. I come every day, on time. I ask for any time off I need in advance. I have gotten wonderful evaluations on my performance. In essence I have become the employee that I have always wanted to be. The money is better then what I was making in the hospital and things have stabilized financially. We still have rocky times, but nothing like it was. Since I am not completely focused on my safety (or lack thereof) and my feeling of overriding failure, I have been able to put time and energy into those things that truly matter in life: my wife, my relationships, my horses, and my growth.

I have been able to get ahead of my health, and take daily medications for my lungs to help with decreasing the intensity of my yearly bronchitis. I spend a lot of time out in pasture. One of my horses is 27 years old, blind, and has Cushing's. I want his last few years to be comfortable and full of love. He deserves that. The other horse that I had during my time at the hospital was a bad fit for me; the relationship that she and I had only fed my feelings of failure. I re-homed her to an amazing woman who truly gives her what she needs. They are truly beautiful together. Since then I have acquired the horse of my dreams since I was a child, a black Percheron. She is kind, quiet, and full of personality.

I still deal with the end products of the PTSD that was created during my time in the hospital. It took about six months on the new job for me to stop stuttering when I spoke with my new boss. I still break out into a sweat when she asks to speak with me, not her fault. She is kind, firm, and never yells or threatens.

As I look over this book throughout its edits and realize where I was versus where I am now, everything feels surreal. Sometimes I sit at my desk and I think "what happened? How did things get so bad?"

I am finally proud of the person that leaving the hospital has allowed me to become.

Here is the first official evaluation that I receive on my new job. I was so happy that I hugged my boss. It reflects the kind of employee that I have always wanted to be:

"K--- Is an important member of the INF team. K--- Has excellent work ethic and is always eager to assist others when needed. K--- Asks questions for clarification or direction, which will assist her in achieving her annual goals and provide her with an increased knowledge base."

www.ingramcontent.com/pod-product-compliance
Lightning Source LLC
Chambersburg PA
CBHW022128170526
45157CB00004B/1798